ESTIMATING ELECTRICAL POWER DEMAND IN BUILDINGS

STEVEN MCFADYEN

Disclaimer

Although the author and publisher have made every effort to ensure that the information in this book was correct at press time, the author and publisher do not assume and hereby disclaim any liability to any party for any loss, damage, or disruption caused by errors or omissions, whether such errors or omissions result from negligence, accident, or any other cause.

Contents

Preface

Electrical power is the primary source of energy for most buildings. An estimation of how much power will be required for a building project is essential in order to facilitate the design of the necessary electrical distribution systems.

Generally speaking, maximum power demand is the issue of most concern to the electrical engineer. Maximum demand within buildings occurs infrequently because normal operations are typically below maximum demand. However, electrical distribution systems must be designed to accommodate maximum demand.

Much of this book addresses the estimation of maximum demand, but it also provides a clear overall understanding of the process of estimating electrical demand. The book offers clear and concise coverage of theory and well-establish procedures — such as IEC methodology and the use of power densities — and concepts are illustrated with practical advice and examples. The techniques can be applied to achieve a realistic estimate of power demand within a building, and the accompanying spreadsheet provides a framework to enable quick and accurate estimation of power demand during early stages of a project.

Current building design is focused on sustainability, environmental impacts, and maximising energy efficiency. All these factors directly affect electrical distribution within a building. In practice, this means that the electrical engineer must collaborate with architects, mechanical engineers, and other parties to achieve the most accurate estimate of demand. It is not uncommon for misunderstandings to arise between project teams in relation to their respective roles in estimating demand. In order to bridge these misunderstandings, this book addresses the limitations of the electrical engineer in these situations and ties together the role of each stakeholder in determining the best power demand estimate.

This book provides a foundation for anyone wishing to develop more comprehensive and sustainable electrical distribution systems for buildings. As such, it is a useful resource not only for electrical engineers, but for building owners, architects, mechanical engineers, sustainability engineers, and others.

The Need to Estimate Electrical Demand

Electrical distribution systems deliver the energy needed to drive the services, systems, and processes within buildings. The estimated electrical demand identifies the requirements of a building in respect of the power supply and distribution. Hence, correct estimation of how much power is required for a building to function successfully is the essential first step in planning and designing the electrical distribution system.

Electrical power is the primary source of energy for most buildings, however, it can be used along with other sources of energy, e.g., gas, fuel, oil, etc. In estimating and understanding the electrical power requirements of a building, the electrical engineer is assessing and understanding the energy consumption of the building, or part of the energy consumption if other energy sources are also used. As such, electrical power and energy are related.

Estimating demand and effective communication

When engineers are asked about power demand estimation, planning of the electrical infrastructure is probably the first thing to come to mind. However, modern sustainability practices employ design and construction techniques to limit energy consumption. This means electrical engineers now have additional data from multiple sources to consider when estimating electrical demand.

If a building uses energy from more than one source, or when sustainability practices limit energy consumption, electrical demand estimation becomes a team effort that involves all disciplines. In these situations, misunderstandings may occur when there is a divergence of views amongst the project teams in respect of their roles. It is usually recognized that estimating demand in these situations should be a team endeavour, but in reality the electrical engineer is often forced to make assumptions in order to estimate the required power for the building.

There are many stakeholders who should be involved in the demand estimation process. Leaving the electrical engineer to do it on his or her own does not yield the best estimate.

The Need to Estimate Demand

Planning of the electrical infrastructure
Determination of spatial provisions
Local supply authority approvals
Costing and project planning
Communication & interface with third parties

A multitude of stakeholders are involved in the design and construction of modern buildings

The Need to Estimate Demand

Architects and engineers play pivotal roles in determining the energy consumption of a building. The size, orientation, and finishes of the building will affect its heat gain and natural lighting, and heat gain and natural lighting will directly affect the performance of MEP (mechanical, electrical, and public health) systems and the amount of energy they consume. Designs that minimise the energy consumption of MEP systems — through building features, equipment and mechanical plant selection, and system design — will have lower electrical demand requirements.

Developers, landlords, and end users also have roles to play in determining the energy consumption of a building. A developer may wish to implement an energy-reduction program across all developments. A landlord may have sustainability and cost goals that will inform the power demand, or the landlord may want to market a certain power level to potential tenants. And it is always necessary to understand the end users' needs when making demand estimates, i.e., an office building intended primarily to be used by banking clients will have larger power demands than a general office building.

Local supply authorities that provide power to the building are also concerned with the demand estimates. Having an agreement with the local supply authority that covers the power supply to the building, anticipated demand, and the configuration of the incoming utilities will ensure that the project runs smoothly and that there are no surprises. In addition, reducing electrical demand can often result in savings on supply authority connection and consumption charges.

When putting together the power estimate for a building, the electrical engineer should be seen as a compiler of input from the various disciplines, not as the sole arbiter of knowledge. As such, it is imperative that effective communication be established between the electrical engineer, project teams working on the design and construction of the building, and other stakeholders during the early stages of a project. This communication is essential to ensuring the electrical engineer makes the best possible power demand estimates.

Benefits of good demand estimating

The aggregate construction cost estimate of a building project is vital to developers, owners, and landlords. An accurate overall budget estimate enables all the stakeholders to plan and strategise to move the project forward. In determining construction budgets, quantity surveyors and cost consultants will require the details of the proposed electrical infrastructure. Hence, a good electrical power demand estimate is crucial to the success of any building project.

A good demand estimate ensures that the budgeted costs for the electrical infrastructure are realistic. It also identifies the equipment and spatial planning necessary to accommodate the electrical supply and distribution systems. An overestimated demand increases the costs of the electrical supply and distribution systems and the supply authority charges. If demand is underestimated, the functioning of the building will be impaired and additional costs will be incurred. In other words, a good demand estimate prevents budget overruns and enables proper provisions to be made for the electrical infrastructure.

The allocation of rooms to contain the electrical infrastructure is a critical part of the design process. Much of the spatial planning for the main electrical rooms is carried out during the early stages of a project. Having a good estimate of the power demand at this stage ensures that proper provisions will be made to accommodate all the necessary equipment, and that an adequate number of appropriately sized electrical rooms will be provided. Once the architecture becomes more fixed, alterations to any electrical rooms are much more difficult to implement, not least because of the additional costs associated with remedial changes.

The benefits of accurately estimating the electrical demand for a project cannot be stressed enough. In short, bad estimates can lead to additional costs that require more capital investment. And of course, this could delay or halt the project entirely.

Power demand estimates and the electrical demand framework

Finding the power demand estimate is best described as part science and part art. A completely accurate demand estimate is never possible, and, at best, an approximate estimate will be obtained. Understanding this allows project teams to better interpret and use the information at hand, and the aim of the demand calculation exercise becomes that of obtaining an estimate of sufficient accuracy to enable the design to continue and the building to be constructed.

The process to achieve correct demand estimation — in terms of stakeholder involvement — can be thought of as the electrical demand framework.

Estimating Electrical Power Demand in Buildings

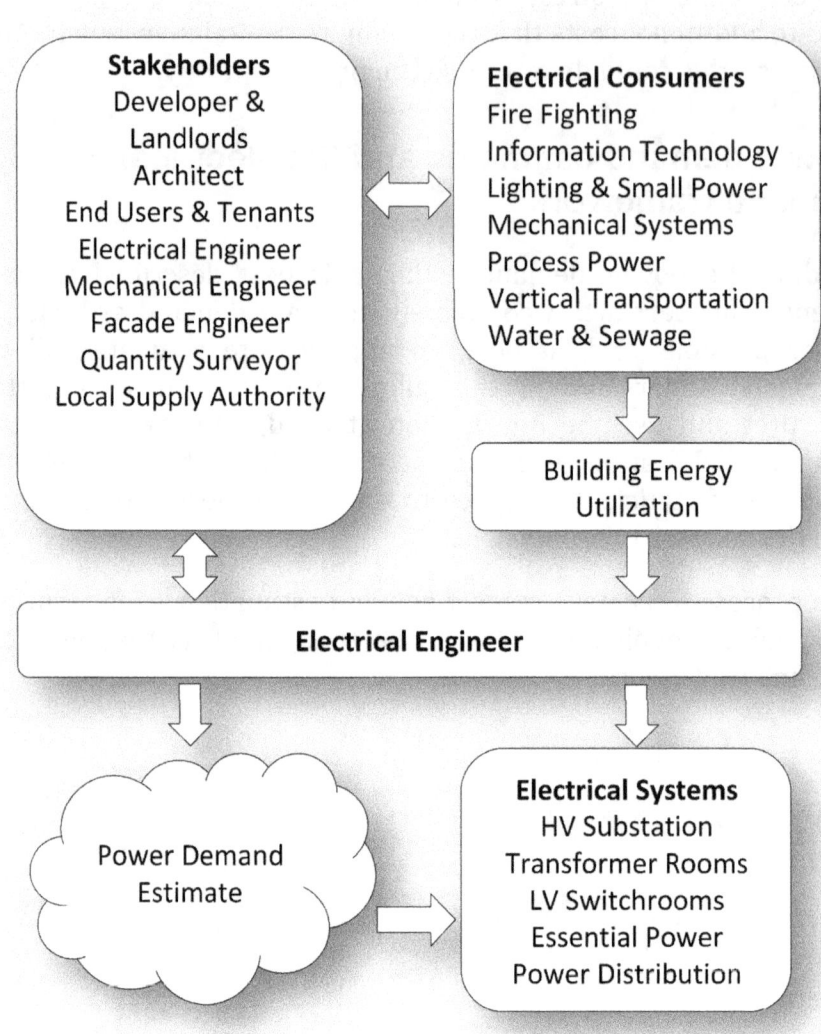

Electrical Demand Framework

The electrical demand framework illustrates the many inputs needed from all the stakeholders and systems in order for the electrical engineer to estimate power demand and plan the electrical infrastructure of a building. The electrical engineer will look at the utility supplies required for the building, any internal substations and transformers, the main electrical switchboards, power distribution mechanisms throughout the building, and any other factors or considerations that will affect the demand estimate and electrical infrastructure.

In pulling the demand estimate together, the electrical engineer is looking to arrive at a figure that gives the team confidence that moving forward utilising the estimate will lead to a functioning design. Having a power demand estimate that not only covers the full building — but one that is broken down for each major area — will allow the electrical infrastructure to be planned in a logical and efficient manner.

Many strategic decisions involving power estimates are taken early on in a project, when there is the least information on which to base estimates. Within this scenario of multiple stakeholder involvement, multiple systems, and limited information, it is important to remember that any accurate assessment of power demand will be a rough estimate at best.

Electrical Power Theory

This chapter is technical in nature. Before looking at the specifics of estimating power demand, it is the view of the author that spending some time reviewing the theory of electrical power is a useful endeavour. It is beyond the scope of this book to explore how the theory was derived, although the reader can easily find this information in any introductory book on electrical engineering.

A better understanding of electrical power theory is not essential in understanding the electrical power demand estimation methods that are presented in this book. However, a working knowledge of electrical power theory will improve the understanding and application of the calculations that are addressed later. This improved understanding will benefit power demand estimation and equipment sizing when planning and designing electrical power infrastructure.

For experienced electrical engineers, this chapter will be a review of power theory. Practitioners new to the electrical engineering field will find the chapter educational, as it will lead to a better understanding of electrical power theory. For non-engineers, the chapter is an opportunity to expand one's knowledge of how electrical power works. While it is recommended that everyone read the chapter, should the reader wish to skip the technicalities of theory, it can be done safely, and the remainder of the book will still be relevant, understandable, and applicable.

What is power?

Power is a measure of the rate of doing work. It is expressed in watts (W). The SI (International System of Units) unit of work is the joule (J). If 200 joules of work are done, 200 joules of energy are expended. The SI unit of power is the watt, and it is equivalent to joules per second (J.s-1).

Note: 1 watt = 1 joule of work per second.

Larger power ranges can be expressed as kilowatts (1 kW = 1,000 W) or megawatts (1 MW = 1,000,000 W).

The unit watt was named after the Scottish engineer James Watt who was famous for his work on steam engines.

James Watt
19 January 1736 – 19 August 1819

Energy is the capacity to do work. When we do something, e.g., moving a box or running at the gym, we are doing work. In physics, this is equated to the energy used in transferring a force through a distance in the direction of the force.

Power measures the rate at which work is carried out (or energy consumed per unit time).

Example: a force of 20 Newtons (N) is used to push an object 10 metres. The object is pushed at a constant speed and it takes 20 seconds to move the 10 metres. How much work is carried out and what is the required power?

— work is 20 Newtons x 10 metres = 200 joules

— required power is 200 joules / 20 seconds = 10 watts

A somewhat common misconception is that watts are an electrical quantity. From the above, it can be seen that this is not true. Watts are a physical quantity that applies to any form of work or energy expenditure.

Vector definition

Our definition of work is the energy used in transferring a force through a distance in the direction of the force. Both the force and distance are vector quantities, and they do not necessarily act in the same direction.

If the force and distance are not acting in the same direction, then we need to take the dot product of the force and distance vectors:

$$Work = \vec{F} \cdot \vec{d} = Fd \cos \theta$$

The angle θ is that between the direction in which the force is applied and that of the movement. If the angle θ is zero, the work is simply the force multiplied by distance. If the force is perpendicular to the distance, cos θ = 0, and no work is done.

Kilowatt-hours

A universal extension of power in electrical engineering is the kilowatt-hour (kWhr). This is simply the power delivered multiplied by the duration. For example, if 1 kW is used for 2 hours, then 2 kWhr are used. As power is the rate of doing work (i.e., joules per second), and this is multiplied by the time (hours or seconds), it can be seen that kWhr is a measure of energy (or work done) rather than power.

With an understanding of the concepts, an example will tie everything together.

Example: a 50 kg weight is lifted vertically 20 metres in 10 seconds by an electric motor (with the lifting force acting in the same direction as the movement). How are the work, power, and kWhr of electricity consumed determined?

— force   = mass x gravitational acceleration
              = 50 x 9.81 = 490.5 N

— work done   = force x distance
                  = 490.5 x 20 = 9,810 joules

— power   = work / time
              = 9,810 / 10 = 981 watts (0.98 kW)

— kWhr   = power (kW) x time (hours)
             = 0.98 x (10/3,600) = 0.0027 kWhr

> Note: as we required kWhr, the number of seconds the motor runs is divided by the number of seconds in one hour (i.e., 10/3,600).

Electrical power

When using electricity to do work, electrical power is measured in watts or kilowatts (W or kW), and it behaves as discussed above. This is often referred to as the active, or real, power (P) of an electrical system.

There is a constantly changing system of electrical and magnetic fields that physically enables the transmission of active power and makes electrical systems work. The need for electrical and magnetic fields to transmit active power is called reactive power (Q). Reactive power is measured in vars or kilovars (VAr or kVAr). Reactive power does not contribute to any work done by a load connected to the electrical system, but it does result in additional current flow and it must be taken into consideration when sizing electrical systems.

The important thing to understand is that active power and reactive power are related by a quantity called apparent power (S). The nice thing about apparent power is that it is simply the voltage multiplied by the current, and the units are VA or kVA.

> Note: apparent power (S) is the product of voltage and current, i.e., volts x amps, hence the unit of VA. This is extended to represent reactive power, Q, by adding an 'r', i.e., VAr.

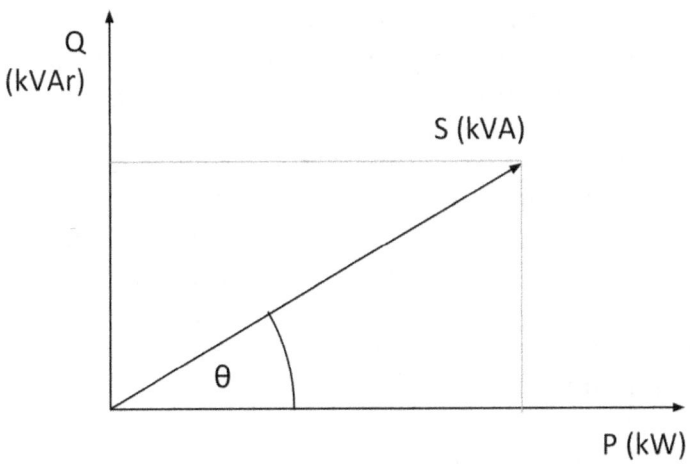

Electric Power

Apparent power (S) is a vector quantity, and it can be broken down into a horizontal component (P, active power) and a vertical component (Q, reactive power), as shown above.

The ratio of active power (P) to apparent power (S) is called the power factor (pf). A useful way to look at it is that the active power used is the apparent power multiplied by the power factor. Mathematically, the power factor can be expressed as the cosine of the angle θ.

From the above diagram, it is possible to see several easy and useful relationships that can be expressed as equations:

$$S^2 = P^2 + Q^2 \text{ giving } S = \sqrt{P^2 + Q^2}$$

$$\text{and } pf = \frac{P}{S} \text{ or } P = pf \times S \text{ and } pf = \cos\theta$$

The power factor varies from 0 (only reactive power present) to 1 (only active power present). As such, it is a direct measure of the amount of reactive power within an electrical system.

Power distribution systems need to be able to generate and/or consume reactive power in the same way that active (real) power is generated or consumed. This requires the installation of reactive generation or consumption equipment, and there are associated costs. Well-designed systems will have a power factor close to 1, thereby minimising the amount of reactive power equipment required.

Summation of electrical power

In electrical engineering, and particularly in the estimation of demand, the power requirements of various items of equipment are usually added together, or summed, to give the total power. Therefore, it is necessary — or at least useful — to understand how power can be summed.

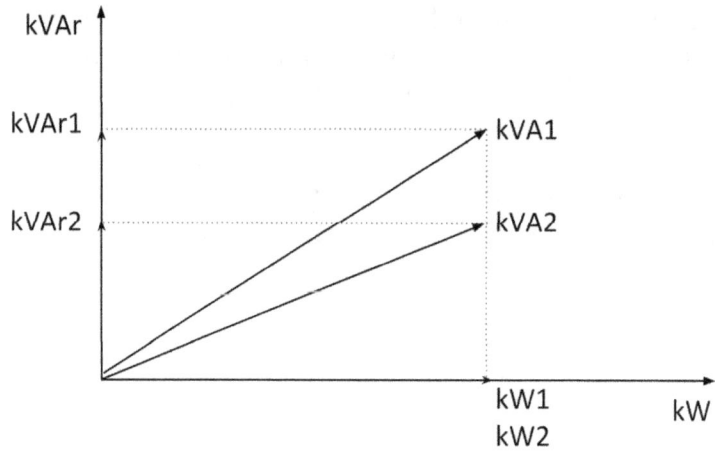

Summation of Electric Power

The above diagram shows two loads of apparent power: kVA1 and kVA2. Bearing in mind the previous discussion and the diagram, it can be seen that apparent power cannot be added arithmetically. Active and reactive power, on the other hand, can be added algebraically. For this reason, it is recommended that power demand estimates be carried out using active power.

Engineers will quite often perform algebraic calculations of apparent power (as is required by some local authorities). While these calculations will invariably contain some margin of error, they will at least be on the safe side of overestimating demand — which is better than underestimating demand).

> Example: assume two loads, kVA1 and kVA2. Load 1 has an active power of 100 kW and reactive power of 90 kVAr. Load 2 has an active power of 220 kW and reactive power of 120 kVAr.
>
> For load 1, the apparent power is 134.5 kVA (see above equations)
>
> For load 2, the apparent power is 250.6 kVA
>
> Total active power = 100 + 200 = 300 kW
>
> Total reactive power = 90 +120 = 210 kVAr
>
> Giving a total apparent power of 366.2 (kVA)
>
> Simply adding the two apparent powers would give 385.1 kVA (with an error of 18.9 kVA).

Complex power

This section is not necessary for understanding the demand estimation methods given later. However, it is presented here to provide a deeper understanding for interested readers, and it will also be of benefit to software developers.

Active (real) power, reactive power, and the power factor can be explained using different concepts. Sometimes, using the concept of complex power is most convenient.

In an electrical system, if the voltage and current are treated as vectors, they are usually expressed as complex numbers. The complex power S (of the system) is then given by:

$$S = V\,I^* \text{ and } S = P + jQ$$

The apparent power is found by multiplying the voltage by the complex conjugate of the current (with both the voltage and current expressed as complex quantities). Using the conjugate of the current results in a positive sign for inductive reactive power and a negative sign for capacitive reactive power.

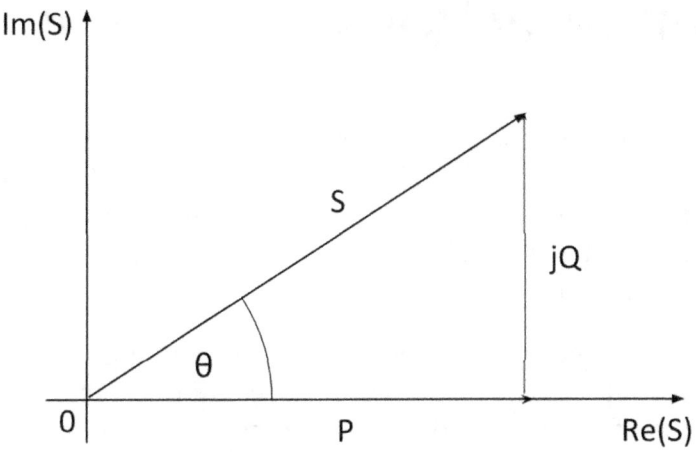

Complex Power

One of the advantages of using complex notation is that consideration of the power factor is inherent in the maths. If you have the ability to add complex numbers in software (or with a calculator), it is easy to add power as complex numbers without sacrificing accuracy.

Estimating Power Demand

The ultimate aim of a demand estimate should be to arrive at a figure with a reasonable level of confidence. The electrical engineer must be comfortable that using the figure in the design will lead to an electrical system that is adequate for purpose without being too big.

There is no single, right answer when estimating power demand, there are only answers that are close enough. Asking several different engineers to estimate the power for a large, complex building will produce several different answers, none of which will match the actual (real) power consumption of the building. The best that can be hoped for from any power demand analysis is a ballpark estimate of the actual power consumption. The good news is that this is generally all that is required.

The problem of demand estimation

Throughout this book, the term 'estimate' is used rather than the term 'calculate' (although they are essentially interchangeable). The primary reason is that it is impossible to determine the power used within a building accurately by calculations. At best, the calculations will give a close approximation.

A few examples are useful in illustrating why it is difficult to estimate power:

> Consider a speculative office core and shell office development. The building electrical supply needs to cater for both the landlord's services and the systems that may be installed by future tenants. At the construction stage, the tenants and equipment they will be using are unknowns.

This makes it impossible to determine any electrical loadings accurately.

Take any item of equipment (air handling unit, water pump, dishwasher, photocopier, etc.), look at the nameplate rating for the electrical power, and then measure the power consumed. It will invariably be less. The reasons for this are numerous — oversizing, selecting the next larger standard sizes, calculation of rating (as opposed to measurement), etc.

At the design stage for large equipment (chillers, air handling units, fire pumps, etc.), power demand will be known, but for smaller devices (dishwashers, photocopiers, computers, light fittings, etc.) power demand will likely need to be estimated.

Most calculations will tend to be on the conservative side. As part of the design process in selecting bulk capacity, the engineer will normally build in a reasonable amount of overcapacity. For example, it is relatively common to apply a 70% load factor at the transformer, thereby giving a 30% buffer (spare capacity).

At all levels — from building intake transformers to final distribution boards — the use of standardised ratings makes it inevitable that systems will have overcapacity. At the same time, the use of standardised ratings will provide a margin of safety in case the power peaks or if power demand calculations are slightly on the low side.

In the author's experience, most electrical systems within buildings have been overdesigned, resulting in too large an infrastructure (transformers, switchboards, etc.) being installed. While it is better to have too much power rather than too little, it should be remembered that there are spatial and sustainability issues — as well as significant costs — associated with overdesign.

Calculating the demand estimate to IEC standards

Estimating power demand is an area of electrical engineering where there is no correct answer; it is a combination of science and art. The electrical engineer plugs the numbers into a preferred method of calculation, and then it is necessary to rely on instincts in order to decide if the answer feels right or not.

Depending on the geographical location of the installation, different methods, numbers, and procedures can be used to estimate the power demand. In this section, we will look at one method in line with what could be considered IEC practice.

To get going, it is useful to understand some basic definitions:

voltage V — the voltage of the electrical system;

load current Ib — the current required to operate an item of equipment;

apparent power kVA — the product of the voltage V and load current Ib;

real power kW — the actual power consumed by the load or equipment;

power factor — the ratio of the real power to apparent power (kW/kVA);

utilisation factor ku — see below;

simultaneity factor ks — see below.

Utilisation factor ku — nameplate ratings invariably list higher values of current than will be seen in use, motors rarely run at full load, etc. A utilisation factor can be applied to these ratings to establish a more realistic load current, thereby not overestimating the demand.

Simultaneity factor ks — not all equipment runs at the same time. For example, one motor may be duty and the other standby. The same applies to installations. For example, a group of houses or apartments will not all consume the full design current at the same time. Applying a simultaneity factor addresses this. The term 'diversity' is often used, and it has the same meaning.

Estimation of power demand is carried out using either apparent or real power. The examples in this chapter use real power, which gives the total actual kW required for the installation.

Estimating Electrical Power Demand in Buildings

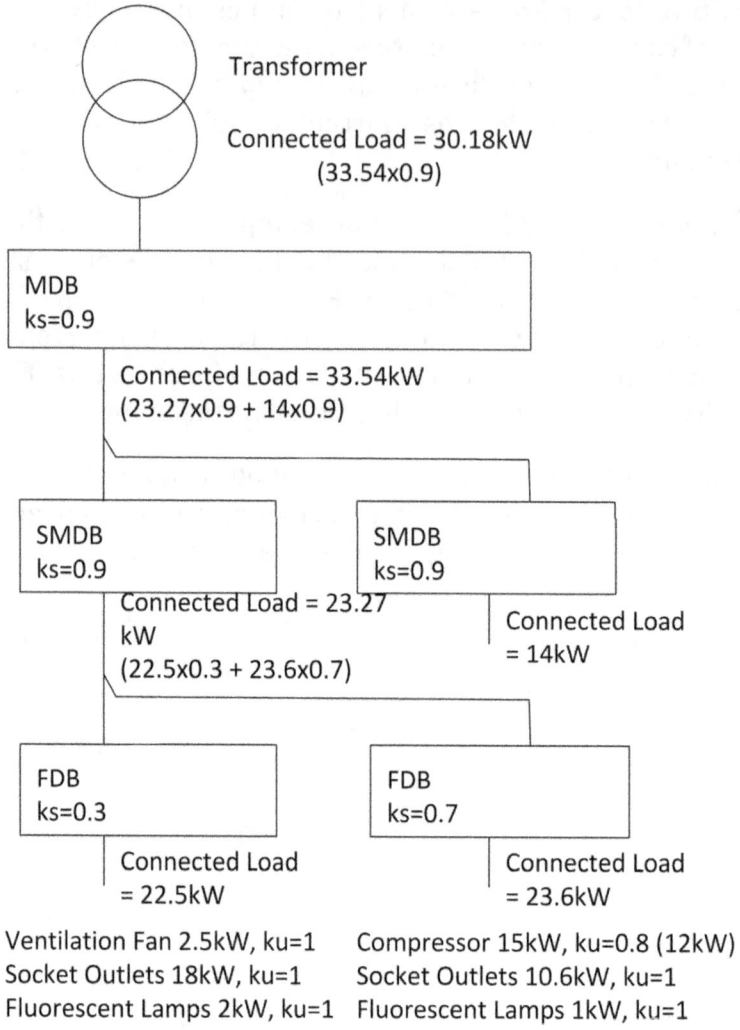

Transformer

Connected Load = 30.18kW
(33.54x0.9)

MDB
ks=0.9

Connected Load = 33.54kW
(23.27x0.9 + 14x0.9)

SMDB
ks=0.9

SMDB
ks=0.9

Connected Load = 23.27 kW
(22.5x0.3 + 23.6x0.7)

Connected Load = 14kW

FDB
ks=0.3

FDB
ks=0.7

Connected Load = 22.5kW

Connected Load = 23.6kW

Ventilation Fan 2.5kW, ku=1   Compressor 15kW, ku=0.8 (12kW)
Socket Outlets 18kW, ku=1   Socket Outlets 10.6kW, ku=1
Fluorescent Lamps 2kW, ku=1   Fluorescent Lamps 1kW, ku=1

IEC Demand Estimation Method

The above diagram illustrates how utilisation and simultaneity factors are used to estimate the power demand of the installation. At the load end, the estimated load demand (nameplate rating or some calculated estimate) is multiplied by the utilisation factor to give a realistic load demand.

> Example: a compressor has a nameplate rating of 15 kW. The electrical engineer and the mechanical engineer decide that the compressor will never work at full load, and a utilisation factor ku of 0.8 would be reasonable. This gives an actual demand to be supplied by the final distribution board (FDB) of 15 x 0.8 = 12 kW.

The connected loads are summed to give a total connected load at the distribution board. Knowing that not all loads will run at the same time, the engineer makes a judgement call about a suitable simultaneity factor. The total connected load is then multiplied by the simultaneity factor to give a demand for the distribution board which is carried upstream (to become the connected load for the upstream distribution board).

> Example: consider three compressors running in N+1 configuration. Only two would ever operate at the same time, with the third being standby in case of failure of one of the other compressors. A suitable simultaneity factor ks, in this case, may be 2/3 = 0.66.

> Example: in a building containing 50 apartments, each apartment will have its own distribution board. Not all apartments will be fully loaded at the same time, and a suitable simultaneity factor will be applied at the main distribution board (MDB) or sub distribution board (SMDB). A simultaneity factor ks of 0.4 may be chosen by the engineer, based on statistical data for similar developments.

The process of calculating the total connected load, multiplying it by a simultaneity factor, and carrying the load up to the next higher level distribution board is repeated until the estimated (diversified) demand at the transformer is found. As a method of estimating power, this technique is straightforward.

The problems that are frequently encountered with this method are in relation to arranging the distribution and then deciding upon simultaneity and utilisation factors. There is considerable debate around which simultaneity (diversity) factors to use, and the answers can vary tremendously.

Deciding which simultaneity factors to use — while ensuring sufficient electrical capacity is provided to the building without vastly oversizing the required infrastructure — involves quite a bit of experience and judgement, in addition to the science. On a design note, standard practice is to size and dimension the distribution board for the total connected load. The diversified demand is carried up to the next higher level distribution board at that board's connected load, with that board then being sized for its total connected load.

In the example in the above diagram, the two FDBs would be designed, sized, and dimensioned to supply loads of at least 22.5 kW and 23.6 kW. The anticipated maximum demand on either board at any one time would be 6.75 kW and 16.52 kW respectively. The MDB would be similarly designed, sized, and dimensioned for a connected load of 33.54 kW, with an expected demand of 30.187 kW.

Example — simple distribution board supplying motors

Assume 6 motors each with 15 kW output power, efficiency 84%, 5 duty, and 1 standby (N+1). After discussion with the team, it is felt that the motors will never operate at full load, and a utilization factor ku of 0.8 would be justified.

— the input power of each motor
    = output power / efficiency
    = 15 / 0.85 = 17.6 kW

— one motor is standby, simultaneity factor ks = 5/6 = 0.83

— total connected load
    = motor kW x number of motors x ku
    = 17.6 x 6 x 0.8 = 84.5 kW

— diversified (estimated) power demand
    = total connected load x ks
    = 84.5 x 0.83 = 70 kW

In applying these results to the system design, it would be reasonable to size the distribution board supplying the motors for the total connected load (84.5 kW), with individual supplies to each motor for the input power (17.6 kW). The demand carried upstream to the next higher level distribution board would be the diversified demand of 70 kW.

Calculating demand — watts per metre squared

The estimate of electrical power is often required early on in a project, before any knowledge of the equipment that will be installed is available. In the case of speculative developments, the building may even be constructed prior to any knowledge becoming available in relation to the equipment that will be installed.

A standard approach to the estimation of power at an early stage is to use a watt per metre squared (W/m2) density estimate. The electrical engineer can multiply the density estimate by the gross floor area and simultaneity factor to obtain a power demand estimate. This is done on an area by area basis, and tables of typical values (including simultaneity factors) are given in the Typical Demand Figures (W/m2) appendix of this book.

Example: consider an office building that consists of 3,000 m2 open-plan office space, 120 m2 lobby space, 320 m2 retail space, and a dedicated chiller for the building is estimated to use 86 kW.

Using tabulated figures given in the Typical Demand Figures (W/m2) appendix, and multiplying by the given simultaneity factors, the electrical engineer can make the following calculations:

— office density = 12 (W/m2) x 0.8 (lighting)
             + 28 x 0.8 (power) = 32 W/m2

— lobby density = 14 (W/m2) x 0.85 (lighting)
             + 1 x 0.3 (power) = 12.2 W/m2

— retail density = 18 (W/m2) x 0.85 (lighting)
             + 42 x 0.6 (power) = 40.5 W/m2

— office power   = 32 (W/m2) x 3,000 (m2)
             = 96,000 W = 96 kW

— lobby power   = 12.2 (W/m2) x 120 (m2)
             = 1,464 W = 1.46 kW

— retail power   = 40.5 (W/m2) x 320 (m2)
             = 12,960 W = 12.96 kW

— chiller power   = 86 kW

Total power required = 96 + 1.46 + 12.96 + 86 = 196 kW

There are some things to note when carrying out this type of calculation. The tabulated W/m2 figures are assumed to exclude significant loads such as chillers, fire pumps, large process loads, etc. In specific instances, other W/m2 figures would be more appropriate. For example, 65 W/m2 is recommended in some grade A office specifications, or the developer may be aiming for particular clients (i.e., banking) and may wish to achieve a specific W/m2. It is also important to recognise that some local authorities and approving bodies may have their own requirements on W/m2 and simultaneity factors.

Estimating demand summary

Every building is unique. The power estimate requires the electrical engineer to have knowledge of the building, its intended use, the owner's requirements, and local regulations and practices. The electrical engineer can use this knowledge along with the methods outlined in this book to determine an estimated power demand with which to progress the building design.

Using the IEC-based approach for demand calculation provides a standardized and consistent way of estimating a building's power demand. Estimates of connected load (particularly early on in the design process) can be achieved by utilising standard W/m2 values. In the absence of anything else, the list of W/m2 figures and simultaneity factors given in the Typical Demand Figures (W/m2) appendix of this book can be used as a starting point.

Considerations

Various methods to estimate electrical power demand have been covered in previous chapters. To utilise these in an efficient way — and to ensure that the most accurate and suitable estimate is achieved — consideration of a number of associated issues should be borne in mind.

Which measure of power, kVA or kW?

The general method of estimating power demand is to sum the individual demands for each item of equipment or area of power consumption to arrive at a total for the building. This can be achieved using either apparent power (kVA) or real power (kW).

Having said that, it must be said that there are differing opinions. The author once attended a presentation on calculating power demand and one of the first slides the presenter pulled up stated that kVA should always be used. The presenter was adamant that this was the correct approach and that using kW was incorrect. On the other hand, the author has worked in several countries where the power supply authorities will not accept calculations in kVA and insist on these being carried out in kW.

Which one to use?

In reality, it may not matter. As previously noted, the power estimation process can be difficult to navigate, but more importantly, the result is not a definitive answer. Rather, it is a ballpark figure that will facilitate moving forward in designing the electrical system.

It does not really matter if kVA or kW is used because the electrical engineer is not looking for a definitive, completely accurate power demand value, but only a reasonable estimate. Alternatively, there may be no choice in relation to the measure of power that should be used. For example, if the power supply authority requires kVA or kW, the calculation needs to be done that way.

The author prefers and recommends that calculations be done in kW. Active power (kW) is measuring the actual power consumed by the electrical equipment, and it is easier to relate to than apparent power. Using kW also allows the electrical engineer to add (sum) different kWs together without introducing any error, while apparent power (kVA) is a vector quantity and carrying out arithmetic will result in errors. Luckily, any calculations utilising kVA will overestimate the power rather than underestimate it.

A bigger picture

Neither kVA or kW by itself will give the full picture. In a power system, the real work done is given by kW, the reactive power by kVAr, and the apparent power by kVA. The relationship between the three measures is the power factor. If the electrical engineer knows the kW (or kVA) and power factor for each item of equipment or area, more mathematically correct calculations can be carried out and the electrical engineer will arrive at a nominally more accurate answer.

The question is whether or not the additional complexity is worth it. The answer is: it depends. It is necessary to bear in mind that the electrical engineer is only looking to arrive at a ballpark figure. Normally, the electrical engineer will be carrying out electrical demand calculations early on in the project. At this stage, the power factor information necessary to carry out detailed types of calculations may not be available. Hence, if the electrical engineer will end up guessing/assuming power factors anyway, is anything to be gained by being super-accurate?

Efficiency

The efficiency of electrical equipment is highest when operating at peak design value. With systems and equipment operating at partial load, the efficiency can drop off quickly.

Typically, overdesigned systems operate at 5% to 10% lower efficiency than needs be. A more appropriately sized system will take advantage of having higher efficiency (less loss), and this will be directly realised in cost savings and improved sustainable design.

Actual equipment ratings

When known, the use of actual equipment ratings will give a better estimate than the use of generic w/m2 figures. It is always worth the effort to obtain this type of information and include it in any estimate. This is particularly important for specialist buildings where the main function is to provide power for process-related equipment.

A useful approach to dealing with actual equipment is to maintain a load list for the project. In addition to helping with the estimate of electrical power, a well-developed load list can provide invaluable assistance in the detailed design of the electrical system.

Typical base-level information to capture on any load list would include designation and description of the load, rated power, power factor, efficiency, utilisation and simultaneity factors, duty (continuous, intermittent, or standby), and supply voltage, phases, and supply type (normal, essential, etc.) for each load. Additional information could be collected depending on requirements (e.g., redundancy information).

Essential (standby) power

In addition to normal power, many services within buildings are backed up by on-site generators in case of utility failure. Along with an estimation of normal power, estimations of essential (standby) power are also required.

The methods already discussed apply equally to the estimation of essential power, and producing these estimates is not difficult. If the W/m2 method is being used with a spreadsheet for example, it is a relatively easy thing to add in columns to identify loads on essential power and sum up these requirements.

When selecting backup generators, consideration needs to be given to both the active power (kW) and the apparent power (kVA). Manufacturers produce operational curves that can be used to select the correct generators.

Transformer sizing

The issue of transformer-sizing merits some consideration. Transformers are invariably specified in kVA. This means that if the power estimate has been completed using kW, it will need to be converted to kVA. Where supply authorities require the power factor to be maintained above some minimum (typically 0.95), this can be used to obtain an estimated kVA.

It is generally not good practice to fully load a transformer based on the demand estimate (for reasons discussed above). Typically, the transformer is selected to be loaded up to a certain percentage (allowing some capacity for future growth or short-term overloading). A load factor of 70% is commonly used, although other figures can be adopted if the engineer so wishes.

> Example: assume the estimated load for a building is 1,390 kW. Using a power factor of 0.95, this is equivalent to 1,464 kVA. With a 70% load factor, this gives 2,092 kVA.

> A good engineering solution could be the installation of two 1,000 kVA transformers. While not quite giving a 70% load factor, it is close, and the use of two transformers instead of one will provide some distribution flexibility.

Issues of distribution and voltage drop play a key role in the positioning of transformers. There will be a large number of transformers on larger buildings and developments, and the locations of the transformers need to be considered carefully.

Having the electrical load estimate broken down into key areas — with sufficient detail to understand the power distribution throughout the building — will greatly assist with the placing of transformers and routing of electrical distribution infrastructure.

In general, building transformers are dry-type, and most manufacturers producing to IEC standards will follow the preferred sizes of 100, 160, 250, 400, 630, 1,000, 1,600 and 2,500 kVA. It should be noted that 2,000 kVA is sometimes used, although this is not a preferred standard size.

Electrical engineers' dilemma

In most design practices, the electrical engineer is given the job of estimating the power demand for the building. Often, the electrical engineer is also given the task of reducing electrical demand (e.g., in order to meet energy efficiency guidelines for a sustainability rating system). However, the reality is:

> *The electrical engineer is often responsible for the determination of the electrical demand, but the factors making up that demand are outside the electrical engineer's control.*

For example, the amount of power consumed by a chiller is dictated by the size of the chiller, the air-conditioning load, and how the chiller is operated. All these are the purview of the mechanical engineer; the power consumed has little to do with the electrical engineer. While this is fairly obvious to many, in the author's experience, a disproportionately large percentage of people seem not to recognise this.

It would appear that sometimes there is an assumption that the electrical engineer can determine the electrical power requirements via utilising personal knowledge. By now, the reader should be aware that many other stakeholders have a direct input in determining how much power a given building will require.

The electrical engineer does not control the amount of power consumed by the building. The role of the electrical engineer is to move the required power from the supply authority to the consuming equipment. If the requirements of a cost-effective good design and sustainable electrical infrastructure are to be met, then clearly, a team approach must be taken.

Spreadsheet

The spreadsheet that accompanies this book provides an easy way to use the watts per metre squared method of estimating demand. The spreadsheet can be used as is, or it can be modified to meet the requirements of any particular project.

Free spreadsheet

Purchase of this book includes a free copy of the myElectrical.com Load Estimate Spreadsheet.

To obtain the spreadsheet, please visit http://myelectrical.com and browse the store. The spreadsheet can be found in the software category. Enter and apply the coupon code 'dmBk100' to receive a 100% discount. After finishing the checkout procedure, the spreadsheet can be downloaded from the digital download locker.

Please use the support options on the site if you experience any difficulties.

Spreadsheet — applying the method

Internally, the spreadsheet has two tabs: 'Power Densities' and 'Load Estimate'. The power densities tab contains lighting and power densities (w/m2 and simultaneity factor) arranged by area type. Default density figures are based on the recommendations of ASHRAE 90.1:2007 and the 2006 Siemens' Application Manual – Basic Data and Preliminary Planning. The density table has been given a defined name — '_SpaceLoads' — which is referenced in the load estimate.

The default densities can be used as they are, and the table can be modified. For example, in the table, you can add in areas not already covered and amend the different load densities — as agreed or required by your project. When modifying the densities, the only thing to keep in mind is that the defined table name should encompass the full table, otherwise the lookup formulae on the load estimate may not work.

The load estimate is designed to be self-explanatory. Different coloured areas represent input required and calculated results. It should be fairly easy to understand the functioning of each column, although the following may be useful:

Level — entered by user — the level of the building (G, 1, ground, first floor, etc.).

Area Assignment Description — entered by user — description of the area under consideration (marketing department, staff canteen, etc.).

Area Assignment Space Type — entered by user — space type selected from dropdown list. The drop down list is automatically updated to reflect the areas entered into the densities tab.

Qty. — entered by user — this can make the entry of several identical areas easier (often it is just left at 1).

Area Unit — entered by user — the m2 of the area assignment under consideration.

Area Total — calculated by spreadsheet — the total area of the assignment under consideration (the product of Qty. and Area Unit).

Lighting (W/m2 and SF) — calculated by spreadsheet — values for the area assignment under consideration should be looked up in the density table tab.

Power (W/m2 and SF) — calculated by spreadsheet — values for the area assignment under consideration should be looked up in the density table tab.

Major Plant Load Description — entered by user — items of major plant (e.g., chillers) that are not covered by the W/m2 densities.

Major Plant (W and SF) — entered by user — power requirements and simultaneity factor for major plant. Note: this is entered in watts to be consistent with other columns.

Power Usage Total — calculated by spreadsheet — total power usage for the area assignment under consideration, taking into account simultaneity.

Power Usage Effective — calculated by spreadsheet — total power usage for the area assignment under consideration, taking into account simultaneity.

Area, total power, and effective power are summed to give the building estimate. Power is converted to kW at this stage to deal with more realistic quantities.

Based on a user input of transformer size, power factor, and load level, estimation is given for the quantity of transformers required for larger buildings. This should only be used as guidance, and the engineer should use judgement as to the best transformer arrangement for the building.

Implied diversity, total W/m2, and diversified W/m2 are calculated for the full building. These can be used to compare against similar buildings as a reality check.

It should be noted that the load estimate is filterable and sortable by any column. This is particularly useful for dealing with large buildings. To get the most out of this feature, each row should be independent (i.e., it should stand alone as an entry and not rely on surrounding rows to provide context). This is easy to achieve, but needs to be kept in mind.

Example calculation — mixed-use development

A speculator wants to build a large mixed-use development consisting of retail, offices, warehouse, parking, and external areas. The total building gross indoor floor area is approximately 45,000m2 and the external area is 9,000m2. The development footprint is approximately 19,000m2 and is to be sited on a plot of 120,000m2 with a significant external landscape area. Based on previous experience, the speculator wants to market the office areas as having an installed power density of 60 VA/m2.

As part of the feasibility study, it is necessary to determine an estimate of electrical power in order to enable the electrical systems to be provisionally dimensioned, and to inform part of the cost analysis.

The initial setup of the spreadsheet shows the calculation for this example in detail. The area assignments reflect the architect's initial vision. In addition, the mechanical engineer has given the estimated power loading for major plant, including chillers, air handling equipment, fire pumps, and water pumps.

The densities sheet has been modified to add in the speculator's requirement of 60 VA/m2 (57 W/m2 at 0.9 power factor) for office spaces, which is higher than the standard default values.

Estimating Electrical Power Demand in Buildings

Space	Lighting		Power	
	(W/m2)	SF	(W/m2)	SF2
example_office_60VA	12	0.85	57	0.8

The results of the calculation show a total connected load of 3,600 kW and maximum demand of 2,700 kW. To satisfy this load, three 1,600 kVA transformers are proposed.

Note: due to continuing development, the spreadsheet may contain minor differences from the details given above.

Electrical Power Energy Targets — An Alternative Method

Estimating power demand for a building early on in the project is a confusing and often debated issue. Most electrical engineers involved in working out how much power to supply will say this is the most difficult part of their job. And if you ask two electrical engineers to tell you the required power for the building, you will likely get two different answers.

The traditional approach is to use W/m2 figures to make the initial demand estimate, and as better information becomes available during the design process the demand is refined. This methodology often leads to oversizing, resulting in the installation of significantly more equipment than is actually required. In addition to increased costs, this approach does not reflect sustainable design.

Despite the fact that reasonable power demand estimation requires a team effort, it is not uncommon for the electrical engineer to be asked to estimate the demand. Often, the discipline engineers responsible for the end user equipment may not care about, or be interested in, the electrical requirements because it is assumed that the electrical engineer will sort them out. In reality, demand is governed by the end user and equipment (mechanical plant, water and drainage, process equipment, etc.), and the electrical engineer has limited control over these things.

In the author's experience, the loading on transformers in most buildings tends to be at the low end. An example of a particularly bad project is a shopping mall for which the author conducted an energy usage audit. The mall had a total installed transformer capacity of 33 MVA and actual measured peak demand of 4.2 MW.

It is not uncommon for human nature, watts per meter squared, and diversity to combine to ensure that power demand will invariably be overestimated. Human nature says it is better to have too much power than not enough. And while this is true, it also leads to an inefficient, costly, and environmentally unsustainable installation.

Is there a way to resolve these issues, or a better approach to estimating energy demand? The answer is: yes. All the electrical engineer has to do is take a different approach. By thinking of the electrical power supplied to the building as a measure of the energy efficiency and carbon footprint, and setting efficiency targets, the electrical engineer can plan and design better, more efficient, and sustainable electrical systems.

Within a building, the electrical power is consumed by many systems:

- Mechanical heating, cooling, and ventilation;

- Lighting systems;

- Information and computer technology systems;

- Vertical transportation;

- Water, sewage, and other pumping systems;

- Fire fighting and other life-safety systems;

- Process and operational equipment.

With the traditional approach to demand estimating, the designers of each system develop their designs according to their own requirements (e.g., maintaining certain temperatures or flow rates for the HVAC), and then they tell the electrical engineer how much power their designs require. In these situations, the efforts toward minimising electrical power consumption are usually minimal at best. How often has a mechanical engineer considered increasing set points to reduce electrical consumption at peak times? How often has a water services engineer scheduled pumps to operate in sequence at times when electrical usage is low? How often do vertical transportation engineers alter the speed/timing of lifts to reduce electrical use at peak times? Unfortunately, the answer is: very rarely.

In the author's experience, there is a better approach to demand estimating. If, at the start of a project, the system designers agree on electrical targets for each system, and the designers are motivated to achieve the targets, this encourages a more integrated design process and leads to the use of less electrical power. Focusing on efficiency and sustainability targets from the outset encourages systems designers to investigate ideas such as those mentioned above — and other more innovative solutions — to the overall benefit of the building.

> Note: Make the energy (power) budget of each system a primary key performance indicator (KPI). Give this KPI the same status and focus as other, more traditional, KPIs (e.g., lighting levels, required ambient conditions, etc.).

The setting of energy targets requires innovative thinking and an integrated team approach that ideally includes all the stakeholders. Tools such as building energy modelling will help in the development of budgets. Setting energy targets (average kWhr and peak kW levels) for each system at the start of a project — and having the responsible engineers commit to designing to meet the targets — will invariably result in a more energy-efficient and cost-effective building.

Some leading and innovative design companies already practice similar design methodologies, and the author has worked on projects where this approach has been utilised to great effect. Not only does it result in a more energy-efficient and cost-effective building, the approach generates innovative solutions for meeting the demand targets of the various systems while producing a more efficient, simplified electrical system.

Energy Demand

The electrical engineer is primarily interested in sizing and installing an electrical infrastructure system that is capable of supplying the necessary power and accommodating maximum power demand for a building. In reality, it is unlikely that the estimated maximum demand will be drawn, and in all likelihood the average demand will be significantly less.

The actual power demand of a building is dependent on time, and it will vary throughout the day. Commercial buildings will consume more power during working hours, with peak demands early in the afternoon. Residential demands typically peak in the early evening.

Typically, within buildings, the consumption of electrical energy is measured in kWhr (kW-hours). The SI unit of power is the watt (W) and the SI unit of energy is the joule (J). To get a feeling of magnitude, 1 kWhr is equal to 3,600 kJ. If the electrical power consumed in kW is multiplied by the duration, this will give the energy consumed by the building. To be more mathematically precise, the electrical power (P) is integrated over time as it is consumed.

$$\text{energy} = \int P \, dt$$

Revenue meters are commonly used to measure electrical energy consumption and provide readings in kWhr. The use of revenue meters for measurement and billing of electrical energy means that there is a vast amount of actual data available — far more data than is available for instantaneous building power demands. With the current emphasis on sustainable design, many countries now have extensive databases of typical energy consumption for various types of buildings. By utilising this newly available information, energy modelling, and energy targets, the electrical engineer can improve energy efficiency and enable the design of more cost-effective electrical distribution systems.

Sustainability experts and building owners already have a strong focus on energy efficiency. This focus — and the associated methodology described above — is gaining traction, although it will probably still be some time before it is widely applied by discipline engineers. However, the electrical engineer can play a significant role in developing these new approaches that are necessary if we wish to build a sustainable future.

Appendix - Typical Demand Figures (W/m2)

The following are typical demand figures for various types of building areas.

Space	Lighting		Power	
	W/m2	SF	W/m2	SF
Atrium – First Three Floors	6	0.85	9	0.3
Atrium – Floors (above first 3)	2	0.85	9	0.3
Audience/Seating Area – Convention Centre	8	0.85	5	0.3
Audience/Seating Area – Exercise Centre	3	0.85	5	0.3
Audience/Seating Area – General	10	0.85	5	0.3
Audience/Seating Area – Gymnasium	4	0.85	5	0.3
Audience/Seating Area – Motion Picture Theatre	13	0.85	5	0.3
Audience/Seating Area – Penitentiary	8	0.85	5	0.3
Audience/Seating Area – Performing	28	0.85	5	0.3

Estimating Electrical Power Demand in Buildings

	Lighting		Power	
Space	**W/m2**	**SF**	**W/m2**	**SF**
Arts Theatre				
Audience/Seating Area – Religious Buildings	18	0.85	5	0.3
Audience/Seating Area – Sports Arena	4	0.85	5	0.3
Audience/Seating Area – Transportation	5	0.85	5	0.3
Automotive – Service/Repair	8	0.85	72	0.6
Bank/Office – Banking Activity Area	16	0.85	54	0.6
Classroom/Lecture/Training – General	15	0.85	15	0.6
Classroom/Lecture/Training – Penitentiary	14	0.85	15	0.6
Conference/Meeting/Multipurpose	14	0.85	26	0.8
Convention Centre – Exhibit Space	14	0.85	26	0.8
Corridor/Transition – General	5	0.85	10	0.3
Corridor/Transition – Hospital	11	0.85	10	0.3
Corridor/Transition – Manufacturing Facility	5	0.85	10	0.3
Courthouse – Confinement Cells	10	0.85	0	0

Space	Lighting		Power	
	W/m2	SF	W/m2	SF
Courthouse – Courtroom	20	0.85	20	0.8
Courthouse – Judges' Chambers	14	0.85	20	0.8
Dining Area – Bar Lounge/Leisure Dining	15	0.85	10	0.3
Dining Area – Family Dining	23	0.85	10	0.3
Dining Area – General	10	0.85	10	0.3
Dining Area – Hotel	14	0.85	10	0.3
Dining Area – Motel	13	0.85	10	0.3
Dining Area – Penitentiary	14	0.85	10	0.3
Dormitory – Living Quarters	12	0.85	18	0.6
Dressing/Locker/Fitting Room	6	0.85	10	0.3
Electrical/Mechanical	16	0.85	15	0.3
Exterior – Stairways	10.8	0.85	0	0
Exterior – Building Facades (illuminated)	2.2	0.85	0	0
Exterior – Building Facades (non-illuminated)	0	0.85	0	0
Exterior – Canopies/Overhangs	13.5	0.85	0	0
Exterior – Entrances/Inspection	13.5	0.85	0	0

Estimating Electrical Power Demand in Buildings

	Lighting		Power	
Space	**W/m2**	**SF**	**W/m2**	**SF**
Station				
Exterior – Loading Areas	5.4	0.85	0	0
Exterior – Sales Areas	5.4	0.85	0	0
Exterior – Uncovered Parking Area	1.6	0.85	0	0
Exterior – Walkways (< 3m wide)	9.9	0.85	0	0
Exterior – Walkways (≥3 m wide)	2.2	0.85	0	0
Fire Stations – Engine Room	9	0.85	15	0.3
Fire Stations – Sleeping Quarters	3	0.85	18	0.6
Food Preparation	13	0.85	387	0.8
Gymnasium/Exercise Centre – Exercise Area	10	0.85	20	0.6
Gymnasium/Exercise Centre – Playing Area	15	0.85	15	0.6
Hospital – Emergency	29	0.85	360	0.6
Hospital – Exam/Treatment	16	0.85	360	0.6
Hospital – Laundry – Washing	6	0.85	360	0.6
Hospital – Medical Supply	15	0.85	360	0.6
Hospital – Nursery	6	0.85	360	0.6

Space	Lighting		Power	
	W/m2	SF	W/m2	SF
Hospital – Nurses' Station	11	0.85	360	0.6
Hospital – Operating Room	24	0.85	360	0.6
Hospital – Patient Room	8	0.85	360	0.6
Hospital – Pharmacy	13	0.85	360	0.6
Hospital – Physical Therapy	10	0.85	360	0.6
Hospital – Radiology	4	0.85	360	0.6
Hospital – Recovery	9	0.85	360	0.6
Hotel/Motel Guest Rooms	12	0.85	18	0.6
Laboratory	15	0.85	360	0.6
Library – Card File and Cataloguing	12	0.85	38	0.6
Library – Reading Area	13	0.85	32	0.6
Library – Stacks	18	0.85	32	0.6
Lobby – General	14	0.85	1	0.3
Lobby – Hotel	12	0.85	18	1
Lobby – Motion Picture Theatre	12	0.85	18	1
Lobby – Performing Arts Theatre	36	0.85	18	1
Lounge/Recreation – General	13	0.85	2	0.3
Lounge/Recreation – Hospital	9	0.85	6	0.3

Estimating Electrical Power Demand in Buildings

Space	Lighting		Power	
	W/m2	SF	W/m2	SF
Manufacturing – Control Room	5	0.85	72	0.6
Manufacturing – Detailed Manufacturing	23	0.85	72	0.6
Manufacturing – Equipment Room	13	0.85	72	0.6
Manufacturing – High Bay (≥7.6 m)	18	0.85	72	0.6
Manufacturing – Low Bay (<7.6 m)	13	0.85	72	0.6
Museum – General Exhibition	11	0.85	69	0.6
Museum – Restoration	18	0.85	62	0.6
Office – Enclosed	12	0.85	28	0.8
Office – Open Plan	12	0.85	48	0.8
Parking Garage – Garage Area	2	0.85	8	0.6
Penitentiary – Confinement Cells	10	0.85	0	0
Penitentiary – Courtroom	20	0.85	20	0.8
Penitentiary – Judges' Chambers	14	0.85	20	0.8
Police Station – Confinement Cells	10	0.85	0	0
Police Station – Courtroom	20	0.85	20	0.8
Police Station – Judges' Chambers	14	0.85	20	0.8
Post Office – Sorting Area	13	0.85	72	0.6

Space	Lighting		Power	
	W/m2	SF	W/m2	SF
Religious Buildings – Fellowship Hall	10	0.85	5	0.3
Religious Buildings – Worship Pulpit, Choir	26	0.85	5	0.3
Restrooms/Toilets	10	0.85	5	1
Retail Mall Concourse	18	0.85	42	0.6
Sales Area (for accent lighting)	18	0.85	0	0
Sports Arena – Court Sports Area	25	0.85	5	0.6
Sports Arena – Indoor Playing Field Area	15	0.85	5	0.6
Stairs	6	0.85	9	0.3
Storage Active – General	9	0.6	6	0.3
Storage Active – Hospital	10	0.6	6	0.3
Storage Inactive – General	3	0.4	6	0.3
Storage Inactive – Museum	9	0.4	6	0.3
Transportation – Air/Train/Bus Baggage Area	11	0.85	19	1
Transportation – Airport Concourse	6	0.85	24	1
Transportation – Terminal Ticket Counter	16	0.85	14	1

Estimating Electrical Power Demand in Buildings

	Lighting		Power	
Space	W/m2	SF	W/m2	SF
Warehouse – Fine Material Storage	15	0.85	5	0.6
Warehouse – Medium/Bulky Material Storage	10	0.85	5	0.6
Workshop	20	0.85	72	0.6

Data sources

Lighting densities in accordance with ASHRAE 90.1-2007, 'Energy Standard for Building Except Low-Rise Residential Buildings (SI Edition)'.

Power densities derived from Siemens Application Manual – Basic Data and Preliminary Planning, 2006.

The above values are for guidance only. Where possible, values based on actual equipment and/or historical data should be used.

About the Author

Steven is a chartered electrical engineer with over three decades of practical experience working in Europe, Africa, the Middle East, Asia, and New Zealand. He's worked on an extensive range of projects, from residential high-rise buildings to transportation systems to mining operations to power stations and petrol plants. His vast experience means that there are very few electrical systems he has not encountered.

Through his work, Steven gained a real appreciation for mentoring and teaching others in his field. In 2002 he started myElectrical.com, an online space for electrical engineers and students. Steven now runs his own electrical engineering consultancy 'myElectrical Engineering Limited' in the UK and is the owner of myCableEngineering.com, a web based cable engineering application.

Connect with the author online:
http://myelectrical.com/users/steven

www.ingramcontent.com/pod-product-compliance
Lightning Source LLC
Chambersburg PA
CBHW070334190526
45169CB00005B/1886

* 9 7 8 1 5 3 0 0 6 2 2 7 0 *